Dear Parents:

Congratulations! Your child is taking the first steps on an exciting journey. The destination? Independent reading!

STEP INTO READING® will help your child get there. The program offers five steps to reading success. Each step includes fun stories and colorful art or photographs. In addition to original fiction and books with favorite characters, there are Step into Reading Non-Fiction Readers, Phonics Readers and Boxed Sets, Sticker Readers, and Comic Readers—a complete literacy program with something to interest every child.

Learning to Read, Step by Step!

Ready to Read   Preschool–Kindergarten
• big type and easy words • rhyme and rhythm • picture clues
For children who know the alphabet and are eager to begin reading.

Reading with Help   Preschool–Grade 1
• basic vocabulary • short sentences • simple stories
For children who recognize familiar words and sound out new words with help.

Reading on Your Own   Grades 1–3
• engaging characters • easy-to-follow plots • popular topics
For children who are ready to read on their own.

Reading Paragraphs   Grades 2–3
• challenging vocabulary • short paragraphs • exciting stories
For newly independent readers who read simple sentences with confidence.

Ready for Chapters   Grades 2–4
• chapters • longer paragraphs • full-color art
For children who want to take the plunge into chapter books but still like colorful pictures.

STEP INTO READING® is designed to give every child a successful reading experience. The grade levels are only guides; children will progress through the steps at their own speed, developing confidence in their reading.

Remember, a lifetime love of reading starts with a single step!

With thanks to Ken Kostel, Director of Research Communications, Woods Hole Oceanographic Institution for his help in the preparation of this book.

Visit us on the Web!
Seussville.com
StepIntoReading.com
rhcbooks.com

Educators and librarians, for a variety of teaching tools, visit us at RHTeachersLibrarians.com

Library of Congress Cataloging-in-Publication Data is available upon request.
ISBN 978-0-593-30618-5 (trade) — ISBN 978-0-593-30619-2 (lib. bdg.)

Printed in the United States of America
10 9 8 7 6 5 4 3 2 1

Would You, Could You Save the Sea?

With Dr. Seuss's Lorax

by Todd Tarpley

illustrated by Patrick Spaziante

Random House 🏠 New York

I am the Lorax.

I speak for the seas!

The oceans need help!

Can YOU help them, please?

The oceans are home
to SO many fish.
Any size, any shape,
any color you wish!

There also are dolphins!

Lobsters and snails!

Tiny pink jellyfish!

Giant blue whales!

9

Clown fish and coral
and much, much, MUCH more
that we cannot see
when we look from the shore.

11

The air that we breathe?
Why, clean oceans
help make it!
They offer fresh air
to our lungs—
and we take it.

But our oceans are hurting.
They're filling up fast
with plastic and garbage.
How long can they last?

Do you know what becomes
of the straw that you sip?
Does it just disappear
after leaving your lip?

Or the grocery store bag
that you use for one minute,
then throw in the garbage
when nothing is in it?

Is it gone? No, no, NO!

It goes SOMEWHERE, you see.

And from there it can land

with the junk in the sea!

Then it bobs in the sea
with the rest of the rot.
Will it ever be gone?
No—probably not.

Can YOU help the oceans?
You certainly can!
The first thing you need
is a Clean-It-Up Plan!

STEP ONE! Please recycle your metal and plastic! Recycling a can when you can feels fantastic!

STEP TWO! Please reduce!

Use less plastic each day.

You do not need that straw!

Keep that straw far away!

STEP THREE! Please reuse
all the things that you can.
Bottles and bags
must be part of your plan.

Now, visit the ocean!

Play in the sand!

Pick up some litter!

Learn all that you can!

If each of us—EACH OF US—
helps save the sea,
just think what a wonderful
world this will be!